ANCIENT SECRETS
OF THE PYRAMID BUILDERS
REVEALED

BY DAVID A CLAERR

Illustrated by David A. Claerr

All photographs are vintage images in the Public Domain

© 2015 David A. Claerr
All Rights Reserved
No portion of the text or illustrations
may be used without the express permission of the
Author and Illustrator.

The "Fair Use" act does not appy if any monetary profit is made from the use, re-print or any other type of publication of any part of this copyrighted work. This stipulation includes monetized internet blogs.

Throughout the centuries, the gigantic, awe-inspiring Pyramids of Egypt have been a source of wonder and speculation. How did the ancient Egyptians, without the assistance of advanced metallurgy and mechanization, achieve their astounding feats of architectural engineering?

The Egyptian dynastic cultures were some of the earliest civilized societies on earth, and the techniques they employed over 5,000 years ago have largely been lost. Many of the tools and devices they used were constructed of wood and other organic materials that have long since been destroyed through the natural processes of decay and consumption by insects, fungus, bacteria and other biota.

There is also very scant mention in historical records of the early Egyptian's construction techniques. Perhaps one of the earliest descriptions is a brief mention by the Classical Greek historian Herodotus, writing in the fifth century BC, about 1,500 years after the Great Pyramids at Giza were built. Herodotus stated that the enormous stone building blocks of the pyramids were lifted by the use of "short pieces of wood". This description, is rather vague and inconclusive, but it could offer a clue as to the nature of a device actually employed. Herodotus described a device of simple wooden construction, rather than more complicated devices, such as gigantic cranes with extremely long beams for leverage, as many modern speculators have proposed.

In considering the terse description by Herodotus, many engineers, inventors and tinkerers have tried to use short planks, or beams, in variously designed and constructed devices to lift or leverage stone blocks, which usually met with very limited success.

But another vital clue was discovered, at the very dawn of modern archeological science. A Victorian-age British "Egyptologist", found an interesting and enigmatic artifact while studying the contents of an ancient pharaoh's tomb. His full name (in traditional aristocratic British fashion) was Sir William Matthew Flinders Petrie.

In the tomb were miniature models of many types, ranging from small portrait statuettes of the royal family to articles of common every-day use. There were, for example, tiny replicas of food items such as loaves of bread in baskets, dried & salted fish and bushels of grain. There were models of household items such as furniture and cooking utensils. In the collection were also models of the items of commerce and industry, such as scale models of boats, together with their crews. There were also miniatures of tools, implements and devices used in construction and building.

Among these replicas of devices were items that were peculiar in design and construction. Petrie was fascinated by these items because he was unable to determine an obvious use for them. The devices were constructed of two panels of wood, cut into crescent arches, spaced apart and connected by a series of wooden rods or dowels. Petrie surmised that they were used in building and construction because of their placement with other tools and devices used for those purposes. He referred to the devices as "rockers", and consequently, they are often still referred to as "Petrie Rockers". The size of the actual rockers depicted in the miniature is unknown, although they are considered to have been rather large, based on a comparison of the scale of most of the miniatures, which, it must be noted, are not consistent in ratio from one type of model to another, but rather, most are approximate in scale to articles in a related set or collection, often set up much like tiny figures on a miniature stage.

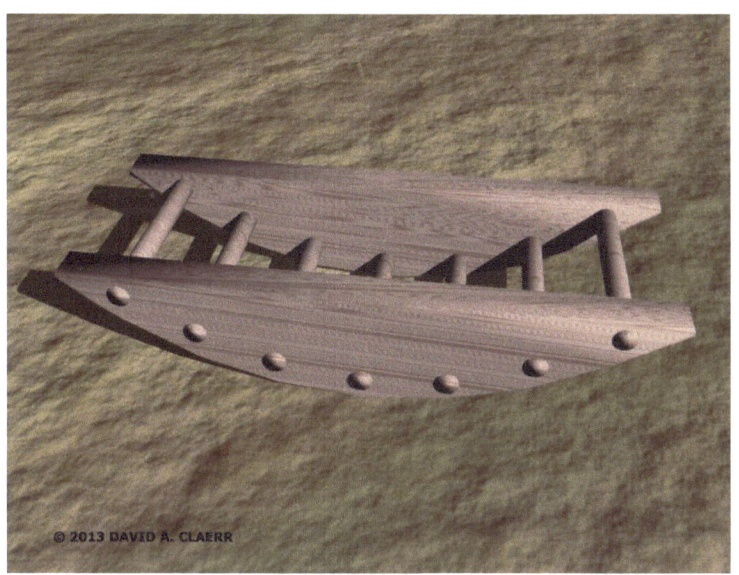

"Petrie Rocker"

Since the time of Petrie's discovery, there have been several attempts to explain or demonstrate the use of the rockers as devices to lift or transport stone blocks. The resulting experiments have had limited success in most cases, and they cannot replicate the lifting of some of the enormous blocks of limestone, sandstone, and in particular, the very dense and heavy granite slabs and blocks that are found in the interiors of the pyramids.

For example, in what is often referred to as the "Kings Chamber" of the Great Pyramid of Khufu at Giza, the walls of the interior chamber and stress-relieving beams above the ceiling are formed by stacks of gigantic granite slabs, some estimated to weigh within the range of 50 to 80 tons!

So, at this point it may be instructive to offer a brief summary of my own experiences and research that can illuminate the use of the Petrie Rockers in the building of the Pyramids.

In my teens and early adulthood, I was involved in various old-school, low-tech wood-harvesting (or lumber-jacking) activities, and the related occupation of constructing log cabins, rough barns and sheds in Michigan, USA and Ontario, Canada. By necessity I became acquainted with the use of a double-bladed ax, crosscut saw, sledgehammer and wood-splitting wedge. These endeavors often required moving large logs and hand-hewn beams with minimal equipment. Often the only motorized equipment that we had was an old pickup truck and a vintage tractor.

To move and position the logs we often used an ancient technique described in medieval literature and surviving wood-engravings, that is often referred to as "parbuckling" The technique uses lengths of rope to move objects. To move a log or beam, we first secured one end of the rope to a stake in the ground, near the place we were to position the log. Next, the rope was coiled around the center of the log or beam in one full circle, then the free end was stretched back to the location of the stake in the ground.

By pulling on the free end, very heavy logs - perhaps 1,000 lbs or more, could be moved by just two persons. The coil of rope around the log provides the same type of multiplied or leveraged force that a pulley mechanism can exert.

Parbuckling a Log

The parbuckling technique worked very well for round logs, but presented more difficulty when moving the hand-hewn beams, which were square in cross-section, and therefore would not roll easily as did the logs with a circular cross-section. Thinking about a solution for this problem, I recalled having seen paintings of carts and wagons from medieval times that had wheels made of solid pieces of wood, cut into crescent arches, and pegged together to complete the circular form. So we adapted this design to some extent, to form detachable wheels around the circumference of the beams, which allowed them to roll as easily as the rounded logs.

When four wooden crescent pieces are attached to an object with a square cross-section- such as those we used when parbuckling large, squared, wooden beams, the combination acts as an axle-and-wheel unit that allows the beams to be rolled easily from place to place, and the beams can even be drawn up inclines using a rope or series of ropes, employed in the parbuckling technique.

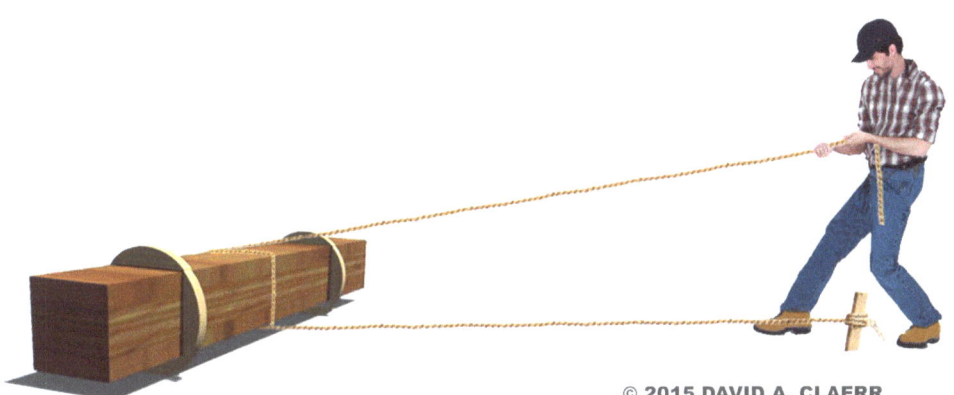

© 2015 DAVID A. CLAERR

Parbuckling a Squared Beam with Wooden Cresents Attached

So, fast-forward to my current era of research. In perusing vintage books on Egyptology, I found various references and some old black-and-white photos of Petries's rockers. When I first saw the photos, I almost instantly recognized the specific crescent shape (technically, an arc formed by the 90 degree intersection of a circle) which makes up one-quarter of the parbuckling wheel's circumference:

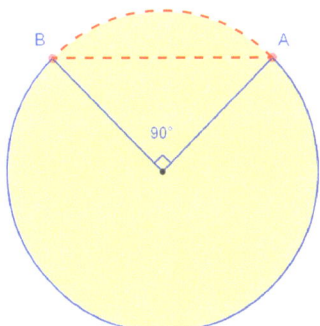

90-degree section of an Arc- the shape of Petrie Rocker sides

If a person has the opportunity to read some of the vintage literature from the Victorian era Egyptologists, it will be noticed that there are many speculations, false assumptions and statements that can easily be disproved, which yet have been passed down and are still repeated to this day.

For example, there are statements to the effect that the pyramids were built "despite the fact" that Egyptians "did not know the use of the wheel" - a notion plainly contradicted by the fact that several pharaoh's two-wheeled war chariots were entombed with them inside of their pyramids.

Although the first recorded introduction of chariots occurs about 500 years after the reign of Khufu, civilizations in Mesopotamia had chariots in use as early as 3,000 BCE., and it is quite likely that the well-traveled scholars and engineers of Khufu's era knew of them. Hieroglyphs and papyrus scrolls also depict wheeled ox-carts nearly identical to the Egyptian cart pictured below, of ancient but currently indeterminate age.

Ancient Egyptian Cart
(Image in Public Domain; circa 1938)

Another theme constantly iterated by the Victorian Egyptologists is the notion that all the dimensions, alignments and fittings are "perfect" "precise" "unerringly accurate" - ideas which were conjured to impress Victorian audiences, but which are easily dispelled when visiting the actual sites.

Even in old aerial photos, now in the public domain, images plainly show the sides of the Great Pyramid to be irregular and out-of-square. (At the center of each side is an irregular furrow or indentation.) The slopes of the sides differ from one another, and the peak or topmost point of the Great Pyramid is offset from the center in relation to the base:

**Offset peak and irregular design of Great Pyramid
(Image in Public Domain)**

Modern, objective scientific surveys using GPS positioning technology also confirm the irregularities mentioned above. The indented sides may have been an part of an intentional engineering design, although they are not visible on Kafre's Pyramid, which is of a similar mass.

It is important to note that the technique using Petrie Rockers that I am detailing was not necessarily used to transport, elevate and position the majority of building blocks used for the pyramids. The "Petrie Wheels", as I will refer to them, were probably used primarily for the largest and most important granite, sandstone and limestone blocks, such as those that comprise the walls, floor and ceiling of the King's Chamber in the Great Pyramid, as well as the elements of the supports and weight-relief superstructure above the chamber.

The Petrie Wheels could fit the descriptions of the "devices made from short planks of wood" that the Egyptian scholars in the time of Herodotus referred to; the specifics of these devices were not detailed by the ancient Greeks.

Another oft-repeated assertion inherited from the Victorians is that "The Egyptians did not know the use of pulleys". Depending on one's definition of a pulley, that may be true in a more-or-less strict modern sense, but at the Great Pyramid there was discovered a type of wooden roller-bearing device that consisted of a hollow wooden roller that fit over a wooden axle, and that functioned as a rudimentary pulley. More information as well as illustrations of this device will be found in the following pages.

It can be easily demonstrated that four Petrie Rockers can be aligned to form a wheel, when attached to an object that is square in cross-section:

Rockers placed to form a circular wheel

Two or more sets of Petrie Rockers were configured around the stone slab to be moved :

Demonstration of Petrie Wheels in place

But how could such a wheel-like configuration be secured to the immense blocks of granite and limestone?

The answer may lie in a study of the Pharaoh's Chariots. The wheels of an Egyptian chariot often consisted of wooden spokes lashed to a circular wooden rim, or tire. The spokes were secured to the rims using lashings of rawhide. Many cultures used rawhide as a fastening material. Rawhide is literally the raw skin of an animal that has not been subjected to a tanning process. The skin has simply had the hair or fur removed and left to dry, usually in the sun. When rawhide is used for lashing, the dry hide is cut into strips. The strips are then soaked in water until soft and pliable. Then the soft strips are wound tightly around two or more objects that need to be fastened together. As the rawhide dries, it shrinks and constricts tightly. When the rawhide is fully dry once again, it becomes exceptionally hard. (For example, Native Americans often set broken bones and used a rawhide cast, much as modern doctors use plaster and bindings to make a hardened cast.) To prevent the hardened rawhide from absorbing water and softening once again, they could be saturated and sealed with either hot grease, hot oil, melted beeswax or a mixture of two or more of these. To remove the hardened rawhide, the lashings could be easily severed with a sharp chisel. Untreated rawhide can be loosened with boiling water.

As an example of the toughness and durability of the rawhide lashings, the lashings on a pharaoh's chariot wheels did not contact the ground, but they were tough enough to withstand the severe torque on the wheels as the chariot raced along over often rough and rocky ground.

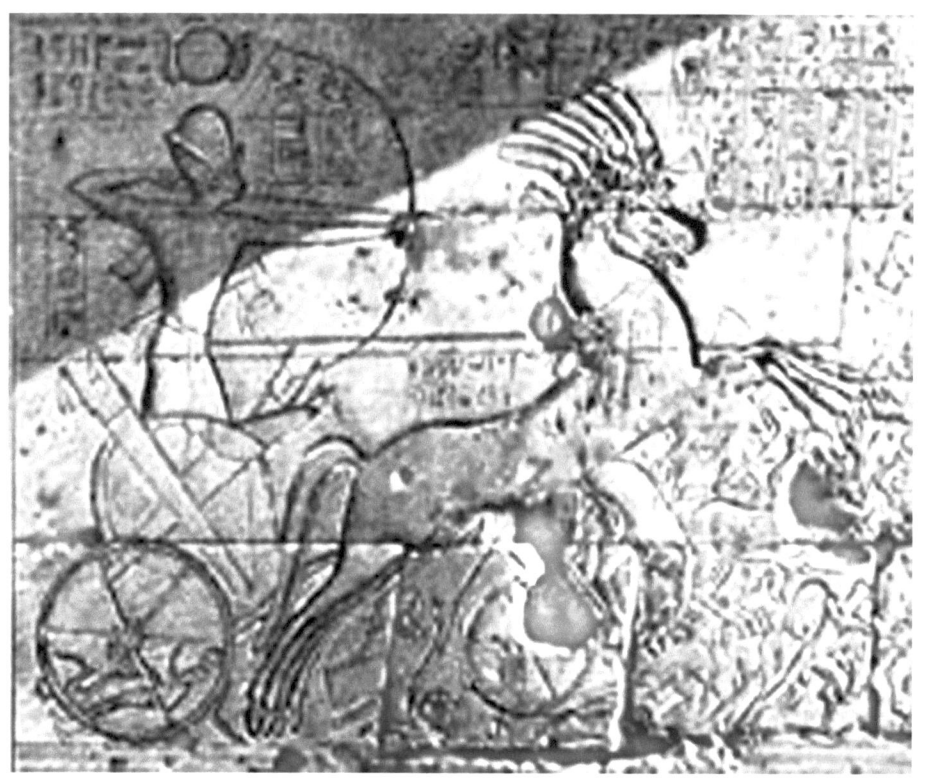

**Chariot of Rameses II
(Image in public domain, photo circa 1887)**

Using rawhide in a related manner, the Petrie Rockers were lashed around the slabs of granite or blocks of sandstone and limestone as demonstrated in the following illustration:

**Demonstration of Petrie Wheels,
set in position and bound with Rawhide**

In this configuration, the Petrie Rockers were firmly bound to the slab and ready to function as wheels or rollers. For exceptionally large stone slabs or blocks, several sets of "Petrie Wheels" were most likely affixed along the slab. Any irregularities in the surface of the slabs could he compensated-for by using folded pieces of wet rawhide, placed under the rockers, against the surface of the slab.

It is thus not difficult to perceive how even the largest 80-ton slabs could be rolled about for transport and positioning when prepared in this manner. But how could they be raised to the heights necessary for stacking and placing, when even many of the large mechanized metal cranes of our current age may not be equal to the task?

To answer this question, a fairly brief summary of the known construction sequences of the pyramids is in order.

There is an often repeated statement (with a few variations), that "the blocks of the pyramid are so precisely cut and fitted that a razor-blade will not fit between the seams, and no mortar or cement was used in the building of the pyramids". This notion is easily dispelled by visiting the pyramids at Giza which, in their present state, have been mostly stripped of their outer casing of mortared limestone, and the exposed surface is now comprised mainly of enormous and very rough-cut blocks, many of which have gaps so large that a person can insert an arm or leg between them.

**The very rough and irregular blocks in the outer levels of Kafre's Pyramid.
(Image in Public Domain)**

And yet, at the top of the "Lesser Pyramid" the tomb of Khafre (or as in the Greek rendering, "Chephren"), there is to this day, at the topmost section, a portion of the original limestone casing still clinging to the peak. (Most of the casing from the pyramids was stripped away for use in the building of the city of Cairo, from many centuries past, up to the 19th century.)

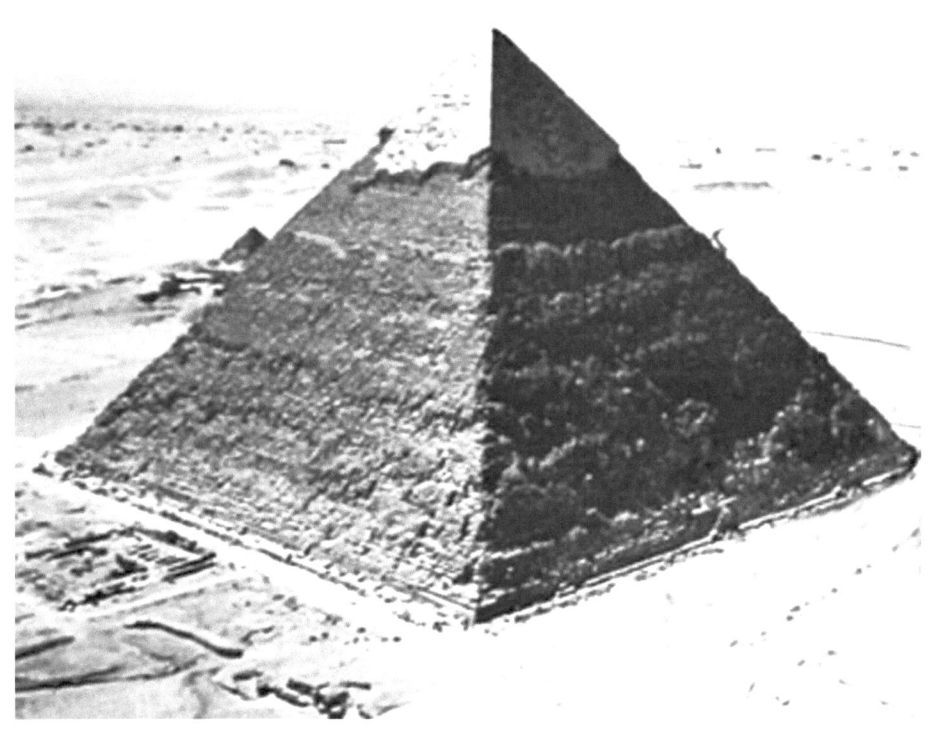

**Pyramid of Kafre with Mortared Limestone Cap.
(Image in Public Domain, circa 1920)**

It is quite obvious that this outer casing is made of a few layers of stonework, with the outermost layers cemented together with a mortar. The mortar has been analyzed and consists of gypsum, quartz, calcite and halite in various proportions. Plaster made of micro-crystalline gypsum was used extensively as mortar on the Pyramid of Khufu as well. The upper reaches of the Pyramid of Khafre are still sheathed by the outermost layers of Tura limestone, cemented in place with mortar. The visual effect of the remnants of the sheathing is that of a cap on Khafre's Pyramid.

If this outer sheath or casing on the summit of the Khafre's Pyramid was not secured by mortar, it would obviously collapse, since it projects several meters from the surface below it.

The pyramids were built in multiple stages:

First, as a base, there was a perimeter of solid blocks with a stepped construction, filled in toward the center with more irregular blocks and rubble, each layered step forming one of a series of stories or layers, with each layer smaller than the previous. During the buildup of these layers, chambers and passageways were constructed on various levels, with previously cut and fitted blocks and slabs of limestone and granite. The process would be similar to erecting small stone buildings and covered passageways on a flat rock foundation. The foundation for the chambers would have been the current layer of the pyramid at the proscribed height. The chambers housed, among other ceremonial items, the sarcophagi of the Pharaohs.

The inner core around these chambers and passageways was often filled with irregular blocks, rubble and fragments, in places cemented together with mortar to stabilize them. In other places, the rubble is simply packed in place, deeper inside from the larger perimeter blocks toward the outer edge of the core. This process followed similar construction methods used in building the mud-brick mastabas, earlier precursors of the stone pyramids.

The perimeter was constructed with the large rough-cut sandstone blocks that were stacked without mortar. The sheer size and weight of the blocks hold them in place.

Then, over the rough shape of the main mass of the pyramid, smaller rough blocks were cut and fitted in a manner that leveled out the surface of the pyramids to a more regular surface.

Over the smaller blocks, another layer of finer white Tura limestone blocks, with surfaces beveled to the slope of the pyramid were fitted. The Tura limestone, cut into interlocking blocks, were cemented with mortar to form the smooth outer layer of the pyramid's skin or surface. The finishing surface blocks were cut with a fair amount of precision and the joints carefully mortared.

There is still today, evidence that the surface of the pyramids may have also been plastered with a mortar stucco containing white gypsum, which gave the pyramids the gleaming white surface, visible for miles, that the ancient written accounts of travelers and adventurers such as Herodotus, often mention.

It is interesting to note that the capstone at the peak of the Great Pyramid was a solid block referred to as a pyramidion, about a meter high and slightly wider at the base. It had the same shape as the pyramid itself. According to Herodotus, the pyramidion was plated with electrum, an alloy of silver and gold.

In the time of Herodotus, as he states, the pyramids at Gisa still had their outer sheathing and a smooth coating of gypsum-mortar stucco as well. The gleaming white exteriors could be seen from many miles away, and the effect of the dazzling sunlight reflected from the silvery electrum plating of the pyramidion capstone was awe-inspiring.

Herodotus not only visited Gisa, between 449 and 430 BC, but also journeyed to nearby Heliopolis, which was the center of learning for Egypt in that era, as it had been for centuries, perhaps millennia. There was a vast library of papyrus texts available. The scholars and priests were still able to be read hieroglyphics and the people of the country still worshiped the Pharaoh as a god. But this was perhaps a few thousand years after the construction of Khufu's pyramid. Although we cannot be certain of the accuracy of his reports, according to Herodotus, he was told by the scholars, concerning Khufu, that:

"Until the death of Sneferu," the priests said, "Egypt was governed excellently and greatly flourished; but after him, his son Khufu succeeded the throne, and he engaged in all manner of wickedness. He closed the all the temples, forbidding the Egyptians to offer sacrifice to the gods. He compelled them instead to labor, one and all, in his service."

Herodotus and other ancient Greek and Roman sources all report that over 300,000 men were employed for 20 years in building the Pyramids. According to these ancient historians, crews of 100,000 men worked shifts of 3 months duration before being replaced by the next shift of 100,000 men.

Although currently there are a multitude of hypothesis concerning who built the pyramids, the length of time that it took, whether or not it was the work of skilled craftsmen or slave labor, these hypothesis are often derived from interpretations of sources such as traditional religious texts, or based on the artifacts and inscriptions recovered from excavations in the vicinity of the pyramids.

Consider first: Why does it have to be an either/or proposition? There was certainly a need for both skilled craftsmen AND the muscle-power of common laborers and a servant class.

In every autocratic monarchy throughout history, and even in parts of democratic America up to the American Civil War, there was an aristocratic bureaucracy supported by a class of commoners and servants or outright slaves, and, as unsettling as that may be to consider, it is true. And it could be true that a supreme monarch such as Khufu who was considered a god, actually did compel nearly every able-bodied person in his kingdom to labor on the pyramids for three months out of every nine, in rotation, leaving six months in-between for each shift to tend to other duties. (Note that the pyramids at Gisa were built at least 1,000 years *before* the biblical narratives of Joseph and Moses, even according to biblical chronology.) So the account related by Herodotus is an important reference to consider, though by no means an established fact. Other hypothesis are essentially unproven at this point in time as well.

That there were slaves in Egypt's extremely long and varied past is a certainty. It is also certain that for periods of time there were Nubian black

Pharaohs over Egypt. But a frieze from the tomb of Horemeb (1321-1293 BCE) shows captives being counted by what appears to be an overseer, at far left. Another Egyptian is seen with a raised club as though ready to strike anyone resisting commands. At right center is a scribe with pen and papyrus in hand, writing a tally of the prisoners. A figure at far right has his arms folded and is cradling his chin with a pensive expression on his face. He holds a cartouche in his right hand, A cartouche is an artifact that functions much like an official's badge- it is an oblong piece, often of ivory or metal, that is inscribed with the royal insignia.

The cartouche shows that the man is an official in the service of the Pharaoh. (Interestingly, the cartouche first appears as an emblem of royalty during the reign of Sneferu, who was the father of Khufu.) The group of officials on this frieze and their activities suggest that the prisoners are to be used for service rather than subject to imprisonment or execution:

Eqyptians enslaved people from many nations that bordered the Mediterranean Sea and Nile Watershed. From the tomb of Horemeb. (Image in Public Domain)

Scenes such as this are not *proof* that slaves alone were used to build the pyramids. Rather they are *evidence* to consider that perhaps part of the workforce was from forced labor. Egyptian conquerors also enslaved people of diverse races and ethnicities from areas all around the southern and eastern Mediterranean basin and Nile watershed. Other evidence, such as graffiti from within the pyramids themselves, suggest that there were also teams of specialized laborers and craftsmen - citizens who took pride in their achievements.

But having a diverse and numerically large workforce would certainly be necessary for projects with the magnitude of the pyramids. Consider that the great cathedrals of Europe often took several generations to complete, and that some remain unfinished today, to varying degrees none-the-less.

According to some estimates, the Great Pyramid of Khufu may contain as many as two-million blocks of stone. However, this estimate is based on the assumption that the entire pyramid is composed of solid blocks, and does not take into account recent discoveries of large amounts of fill comprised of rubble or irregular stone. Ground-penetrating radar indicates that the Great Pyramid is also built over a sizable mound of sandstone that is a natural outcropping of the bedrock. There is also recent evidence in thermal scans of the pyramid that suggest that there may have been a system of internal spiraling ramps that could have been hollow passageways, but the passageways have not actually been verified to date. The Great Pyramid also holds other enigmas, such as a sizable chamber apparently filled only with sand.

The visible external surface of the Great Pyramid, as well as the others at Gisa, show distinct stratified layers. These layers have a consistent composition - that is, each layer has the same grade of sandstone, with a uniform thickness in each layer that by-and-large differs from the layers above and below. Engineers have postulated that the reason for the strata in the pyramids is that the sandstone was removed from the nearby quarries that also have corresponding stratified layers.

At the layer boundaries, the quarried stone separates along the horizontal fracture line of the sedimentary deposits placed down in ancient sea-beds.

The bed layers, built up over hundreds of thousand years, often varied in composition and thickness due to changing climatic conditions such as alternating cycles of dry and wet periods each lasting thousands of years. That is why, as even Sir W.M.F. Petrie, in his very careful surveys of the Great Pyramid, found thicker layers and larger blocks of limestone above thinner layers and smaller blocks, that occurred in random sequences.

During the first phase of constructing the stepped inner core, which would comprise the main bulk of the pyramid, the placement of stone blocks proceeded layer by layer, each layer basically inset at a fairly regular rate, and there would be a stair-step effect, each layer being one step. So it would be relatively easy to have short ramps that simply led from one level or layer to the next.

There is evidence that many of the filler blocks, and some of the outer beveled casing stones were mortared in place in a narrower ascending band to form a solid ramp as the layers progressed. The resulting continuous ramp would be flush with the slope of the pyramid face, and be thus eventually both part of the finished structure, and when the construction was finished, nearly undetectable from the rest of the pyramid face. Remnants of these ramps can be seen today, since they were cemented more firmly than the rest of the casing and therefor more difficult to strip away for building materials.

Note that the each strata of the layers runs across a flat horizontal plane. If the sides of the pyramids had spiral ramps as part of the construction, these sequential horizontal layers would not be evident, and the courses of the stone would follow a spiraling incline around the sides.

**The Stratified layers of the Great Pyramid can be seen on the left. The light-colored, more solid patch of mazonry under the opening at right is possible evidence of a ramp foundation.
(Image in Public Domain, circa 1920)**

Next we will proceed with descriptions of the roller-bearings, a type of rudimentary pulley, which were discovered at Gisa. Also know as "rope rollers", these devices were more sophisticated than that name implies. The bearings consist of a hollow outer cylinder, tapering slightly to the middle, where there is a groove incised to the width of the rope employed. Inside the outer cylinder, is a shaft of wood that is a harder type, which serves as an axle:

Roller-bearings, a type of rudimentary pulley discovered at Gisa

    The roller-bearings of this design were most likely lubricated with a heavy grease. The strong hemp ropes used by the Egyptians would move very smoothly over these bearings with minimal friction. They were mounted in a limestone or granite base, or supporting stanchion, which has recesses shaped to fit the axle ends. Grease on the axle-ends would also cause the entire bearing assembly to rotate with less friction.
    The the recesses of the stone base may have also been highly polished for a glass-like surface, or, replaceable bushings of oiled rawhide may have been fitted to the stone. The wood used for the bearings have been tentatively identified as Lebanon cedar for the outer cylinder, and Nile acacia for the axle.

    The base of this roller-bearing assembly could be mortared firmly in place, yet be removed with relative ease using copper or bronze wedges. (Yes, a bronze tool was found sealed within one of the lower chambers of the Great

Pyramid- indicating that the Egyptians of the time at least had access to bronze. Just as they alloyed gold and silver to make electrum, they may have alloyed copper, nickle and arsenic to make bronze, during the Fourth Dynasty, though evidence of this is rare.)

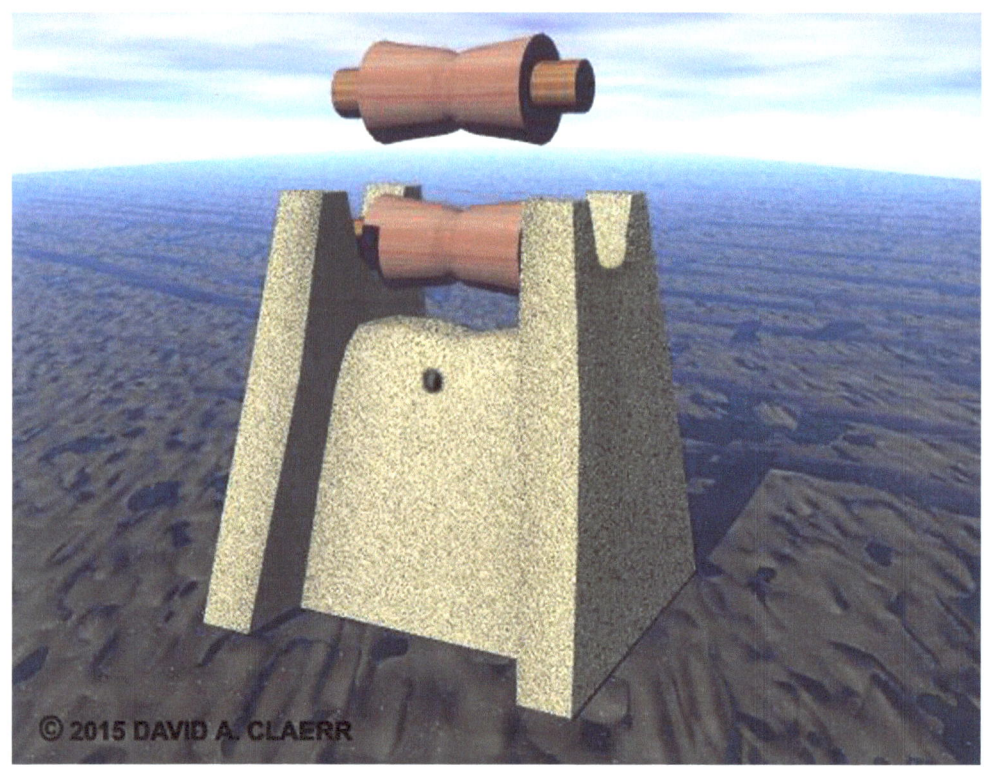

Demonstration of Roller-bearing placed in supporting stancion. One end of the rope could be secured by threading it through the hole in upper center, and fastened with a knot tied in back.

Note that the roller-bearing assembly is not speculation, these are real, tangible objects that the Egyptians used at Giza.

So the combination of the Petrie Wheels, the Roller-bearing Assembly and strong Hemp Ropes were quite likely the means of both transporting stone slabs, and also the method used to raise them up the slopes of the pyramids. This can explain why there is no evidence of stone ramps as in the traditional speculations: the ramps as pictured from the Victorian age until quite recently would entail a mass of material nearly equal to the pyramids themselves. (There are papyrus records of smaller ramps made with a superstructure of wood, and packed with vertically stacked bundles of reeds, that would have been very strong but relatively light. These ramps could have led upward from each layer to the next, and be easily moved higher as the layered strata of the building progressed.)

Explorations into the interior of the pyramids have found significant amounts of packed and sometimes, mortared rubble. It seems evident that each layer of a pyramid was constructed by laying out a perimeter several courses deep of interlocking blocks. Deeper inside this perimeter is the fill of packed and/or mortared rubble. (The exceptions to this were the chambers, which were basically constructed first, as free-standing stone buildings on a flat layer of strata.)

So the use of fill served in this manner: the cutting scraps from the shaping of the formally squared and fitted blocks were both efficiently used and handily disposed of - an ancient recycling process, so to speak.

It also seems to make sense from a practical engineering standpoint that after the stepped layers were completed, it would be advantageous to continuously finish out the outermost skin or casing of the pyramids, on each step or strata, starting with the beveled face stones of fine Tura limestone and progressing outward to the stuccoed surface. In this manner, the beveled face stones could be lowered from above and fit into place. The beveled face stones are stacked atop one another and are cut to interlock with the smaller filler blocks inside the slope face. It is hypothesized that the stucco may have been applied starting from the pyramid top, but that type of process seems more unlikely, given the logistic difficulties it would present.

So the stone slabs were effectively transported, raised and positioned using a relatively simple, yet ingenious combination of devices:

The devices were constructed of mostly native materials, supplemented with common trade items, such as the wood from the famous Cedars of Lebanon, which, as an interesting side note, the royal boats discovered in the pits beside the pyramids at Gisa, were also fashioned from.

For the largest monolithic blocks of granite, sandstone and limestone, multiple Petrie wheels, several individual lengths of rope, and also several Roller-bearing devices were probably employed for each block.

Another type of evidence that Petrie Wheels were used to transport and position some of the more massive blocks is that, around the base of the Pyramids at Gisa, on the plateau above the Nile basin, there is a large leveled area with fairly flat and regular stone paving. Such a stone surface would have created difficulties in moving sledges across, because of the significant friction of wood on stone, but the pavement would be an optimal surface for rolling slabs with Petrie Wheels attached.

Egyptians often transported heavy loads on sledges across wet, packed desert sand for long distances in more-or-less straight lines, but positioning and moving objects on sledges from side to side, or rotating and aligning them, presents more difficulties on a hard, flat surface, even with solid rollers made of logs.

The use of Petrie Wheels would enable large slabs to be rotated and moved in any direction on a hard, flat stone surface with relative ease, and by far fewer laborers.

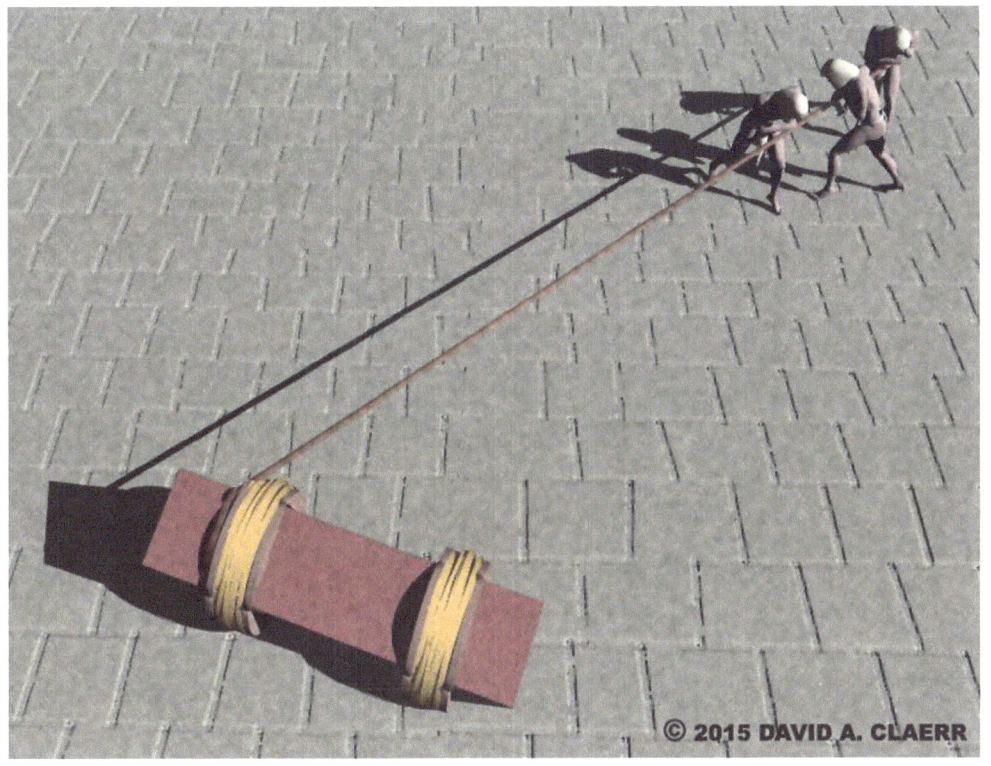

In conclusion, I believe that replicas of these devices can be readily made and demonstrated, either with modern tools, or by using the ancient tools and materials that Egyptians of the era employed. At this present writing, I would encourage interested parties to contact me to collaborate on such an experiment or to notify me if attempts are made independently to test my theory.

LITERARY REFERENCES

Ancient Egypt The Great Discoveries (A Year-by-Year Chronicle)
Reeves, Nicholas
2000
Thames & Hudson, Ltd
ISBN 0-500-05105-4

Atlas of Ancient Egypt
Baines, John; Malek, Jaromir
1980

Les Livres De France
Complete Pyramids, The (Solving the Ancient Mysteries)
Lehner, Mark
1997
Thames and Hudson, Ltd
ISBN 0-500-05084-8

Dictionary of Ancient Egypt, The
Shaw, Ian; Nicholson, Paul
1995
Harry N. Abrams, Inc., Publishers
ISBN 0-8109-3225-3

Encyclopedia of Ancient Egyptian Architecture, The
Arnold, Dieter
2003

Princeton University Press
ISBN 0-691-11488-9
Excavating in Egypt: The Egypt Exploration Society 1882-1982
James, T. G. H.
1982
University of Chicago Press, The
ISBN 0-226-39192-2

Illustrated Guide to the Pyramids, The
Hawass, Zahi; Siliotti, Alberto
2003

American University in Cairo Press, The
ISBN 977 424 825 2

Pyramids, The (The Mystery, Culture, and Science of Egypt's Great Monuments)
Verner, Miroslav
2001
Grove Press
ISBN 0-8021-1703-1

Pyramids of Ancient Egypt, The
Hawass, Zahi A.
1990
Carnegie Museum of Natural History, The
ISBN 0-911239-21-9

Treasures of the Pyramids, The
Hawass, Zahi
2003
American University in Cairo Press, The
ISBN 977 424 798 1